U.S. SPECIAL OPS FORCES

INSIDE MARINE FORCE RECON

HOWARD PHILLIPS

PowerKiDS press

NEW YORK

Published in 2022 by The Rosen Publishing Group, Inc.
29 East 21st Street, New York, NY 10010

First Edition

Editor: Greg Roza
Designer: Rachel Rising

Portions of this work were originally authored by Laura Loria and published as *Marine Force Recon*. All new material in this edition was authored by Howard Phillips.

Photo Credits: Cover U.S. Marine Corps photo by Cpl. Josue Marquez/Released 190716-M-IN847-0003. JPG; cover, pp. 1, 3, 4, 5, 6, 7, 8, 10, 11, 12, 13, 14, 15, 16, 17, 18, 19, 20, 21, 22, 23, 24, 25, 26, 27, 28, 30, 31, 32 CRVL/Shutterstock.com; cover, p. 1 Zsschreiner/Shutterstock.com; cover, pp. 1, 3, 30, 31, 32 Massimo Saivezzo/Shutterstock.com; p. 5 Scott Olson/Stringer/Getty Images News/Getty Images; p. 6 Chris Hondros/Staff/Getty Images News/Getty Images; p. 7, 13, 15, 17, 21, 23, 27 Khvost/Shutterstock.com; p. 7 Christopher Rye/Contributor/Moment/Getty Images; p. 8 Stocktrek Images/Getty Images; p. 9 John Moore/Staff/Getty Images News/Getty Images; pp. 11, 17, 25 David McNew/Staff/Getty Images News/Getty Images; p. 12 https://commons.wikimedia.org/wiki/File:US_Marine_Corps_Force_Reconnaissance_Insignia.svg; p.12 https://commons.wikimedia.org/wiki/File:1st_Recon_Bn_Color.jpg; p. 13 boonchai wedmakawand/Moment/Getty Images; p. 15 Bob Krist/The Image Bank Unreleased/Getty Images; p. 16 José Nicolas/Contributor/Corbis News/Getty Images; p. 19 Hulton Archive/Stringer/Archive Photos/Getty Images; p. 21 Hulton Deutsch/Contributor/Corbis Historical/Getty Images; pp. 23, 28 U.S. Navy/Handout/Getty Images News/Getty Images; p. 27 Joe Raedle/Staff/Getty Images News/Getty Images; p. 29 DMITRY KOSTYUKOV/ Contributor/AFP/Getty Images.

Library of Congress Cataloging-in-Publication Data

Names: Phillips, Howard, 1971- author.
Title: Inside Marine Force Recon / Howard Phillips.
Description: New York : PowerKids Press, [2022] | Series: U.S. Special Ops forces | Includes index.
Identifiers: LCCN 2020036532 | ISBN 9781725328839 (hardcover) | ISBN 9781725328815 (paperback) | ISBN 9781725328822 (6 pack)
Subjects: LCSH: United States. Marine Corps. Force Reconnaissance--History--Juvenile literature. | United States. Marine Corps--Commando troops--Juvenile literature.
Classification: LCC VE23 .P45 2022 | DDC 359.9/84--dc23
LC record available at https://lccn.loc.gov/2020036532

Manufactured in the United States of America

CPSIA Compliance Information: Batch #BSPK22. For further information contact Rosen Publishing, New York, New York at 1-800-237-9932.

CONTENTS

WELCOME TO FORCE RECON

Special "ops" (short for "operations") forces are **elite**-level military units that do jobs no other units can do. Many people have heard of the Green Berets and Navy SEALs, two highly effective U.S. special ops units. Marine Force Recon is another.

Force Recon marines provide essential intelligence to other military units. These specially trained marines travel all over the world to collect information the military couldn't otherwise learn. They find safe places for troops to land and help plan missions. Sometimes, Marine Force Recon carries out its own missions as well, but its main role is to support larger units.

"RECON" IS SHORT FOR "RECONNAISSANCE," WHICH IS A SURVEY OF AN ENEMY REGION TO GAIN INFORMATION. THIS INFORMATION IS REFERRED TO AS "INTELLIGENCE."

MORE THAN THE CORPS

Force Recon marines work differently from members of the regular Marine Corps. A marine follows orders and carries out a plan put together by his commanders. A Force Recon marine is ready for many situations and often has to take action on their own.

Force Recon marines work on covert, or secret, missions and are often needed where larger forces won't work. They're well trained to operate in the air, on land, or at sea. Force Recon marines are also trained to be patient. Sometimes they let enemies get away in order to watch them and learn more about them.

FIRST FEMALE FORCE RECON

FORCE RECON HAS HISTORICALLY BEEN AN ALL-MALE UNIT. IN NOVEMBER 2019, LANCE CORPORAL ALEXA BARTH BECAME THE FIRST WOMAN TO APPLY TO AND COMPLETE THE TRAINING COURSE, MAKING HER THE FIRST FEMALE FORCE RECON MARINE.

THE WORK THAT FORCE RECON MARINES DO PAVES THE WAY FOR OTHER MILITARY TEAMS TO ACT QUICKLY AND EFFICIENTLY, IMPROVING THEIR CHANCES OF SUCCESS.

Many Force Recon teams gather information deep behind enemy lines, and their missions can be very risky. Force Recon commonly works in teams of six marines, though more or fewer may go on certain missions. These small teams allow them to move quickly and quietly through an area while still carrying enough weapons and supplies.

There are three active Force Recon **battalions**, each with their own headquarters. There are one or more Force Recon **reserve** battalions. However, Force Recon operates in secret. There's no way to confirm how many battalions there are today.

FORCE RECON SOLDIERS WORK IN SMALL TEAMS AND OFTEN TRAVEL IN SECRET.

GREEN AND BLACK

Force Recon operations are separated into two types. Green ops are deep-reconnaissance missions, which are mainly for gathering hard-to-find intelligence. These missions rely on the secrecy of small teams. Green ops occur deep behind enemy lines, so each member has to be ready for anything. If the enemy discovers the marines or there are shots fired during a green-op mission, it has failed.

Force Recon marines gather intelligence and then report to their commanders. Types of information gathered during a green op may include mapping roads and waterways and finding good places to land aircraft. Green-op missions prepare other military units for action.

FORCE RECON MARINES HAVE TO BE READY TO WORK IN ALL KINDS OF CONDITIONS.

Black-op missions involve direct action against an enemy. The goal of a black-op mission is to attack and get out quickly. During these missions, Force Recon marines may **seize** enemy supplies, ships, or airfields. Black ops may also involve rescuing prisoners, but Force Recon marines aren't commonly part of those missions.

Black ops are even more secretive than green-op missions. After completing a mission, members of a black-op team must have a plan for quick escape by air, land, or sea. Other military units offer support during black-op missions and help Force Recon units to escape unseen.

MARINE MOTTOS

"SEMPER FIDELIS," WHICH IS LATIN FOR "ALWAYS FAITHFUL," IS THE **MOTTO** OF THE MARINE CORPS. MARINES OFTEN SHORTEN THE SAYING TO "SEMPER FI!" THE MOTTO FOR FORCE RECON IS "SWIFT. SILENT. DEADLY." THIS SOMETIMES APPEARS IN LATIN AS WELL: "CELER. SILENS. MORTALIS."

AMPHIBIOUS RECONNAISSANCE IS AN IMPORTANT PART OF FORCE RECON'S DUTIES. THIS INCLUDES SCUBA DIVING, SCOUT SWIMMING, COASTLINE RECONNAISSANCE, AND SMALL-BOAT OPERATIONS.

FORCE RECON TRAINING

Anyone who wants to become a member of Force Recon must first pass the training to become a U.S. marine. This includes 12 weeks of intense physical exams, history lessons, hand-to-hand combat, water survival, **marksmanship**, and much more.

A marine who wants to join Force Recon must have an excellent military record. They also have to be superior shooters and very smart. Force Recon marines must be even faster and stronger than regular marines. Many marines who apply for Force Recon training don't complete it. In order to be accepted into the training course, marines need to pass a mentally and physically demanding test.

THE PATH TO FORCE RECON

MARINE SPECIAL OPS TRAINING STARTS IN MARINE SPECIAL OPERATIONS SCHOOL. THIS INVOLVES MONTHS OF PHYSICAL CONDITIONING AS WELL AS SPECIALIZED TRAINING, INCLUDING TRAINING FOR BATTLE IN SMALL SPACES, AMPHIBIOUS OPERATIONS, RECONNAISSANCE, WEAPONS, AND MANY OTHER AREAS OF WARFARE AND INTELLIGENCE GATHERING. THOSE WHO PASS MOVE ON TO EVEN MORE SPECIALIZED TRAINING.

MARINE **RECRUITS** LEARN A LOT DURING THE 12-WEEK TRAINING PERIOD. IN ADDITION TO PHYSICAL TESTS, THEY ARE ALSO EXPECTED TO LEARN HOW TO WORK TOGETHER AS A TEAM. THIS IS ESPECIALLY IMPORTANT FOR FORCE RECON MARINES.

Those who are accepted into Force Recon training learn many tasks other marines don't. Training includes physical conditioning on obstacle courses and long runs with heavy packs. Recruits also go through specialized training, such as helicopter rope **suspension**, mission planning, **navigation**, and amphibious reconnaissance.

They also receive training in gathering intelligence. Some of the most important training involves learning to move silently and without being seen. These special marines must be prepared to confront a **hostile** enemy, so they learn skills such as **sniping** as well. Once a marine is accepted into Force Recon, they continue training in highly specialized areas.

RECON TEAM MEMBERS

ONCE MARINES ARE WELL TRAINED, THEY ARE PLACED IN SMALL TEAMS OF AT LEAST SIX. TEAMS INCLUDE AT LEAST ONE MEMBER FOR EACH OF THE FOLLOWING RESPONSIBILITIES: TEAM LEADER, ASSISTANT TEAM LEADER, RADIO OPERATOR, ASSISTANT RADIO OPERATOR, POINT MAN (SOLDIER AT THE FRONT OF THE GROUP), AND SLACK MAN (SOLDIER DIRECTLY BEHIND POINT MAN). ONE MARINE MIGHT TAKE ON SEVERAL ROLES AT ONCE.

THESE MARINES ARE TRAINING IN URBAN WARFARE AT CAMP PENDLETON IN CALIFORNIA.

THE STORY BEGINS

In January 1942—one month after Japanese forces bombed the U.S. military base at Pearl Harbor—the U.S. Marine Corps created the first reconnaissance units. In 1943, these units became the Amphibious Reconnaissance Company (ARC). This battalion included 98 marines. By 1944, ARC had worked with the U.S. Navy on more than 150 reconnaissance missions in the Pacific. They scouted landing locations on Tinian Island, Iwo Jima, and Okinawa.

For the rest of World War II, the battalion continued to assist and train with the navy. They learned underwater operations such as **demolition** and diving. The ARC soon included 20 officers, 270 marines, and 13 navy doctors.

THIS MARINE IS UNDERGOING AMPHIBIOUS TRAINING IN 1945 IN JAPAN.

GATHERING FORCE RECON

In March 1951, ARC was tasked with landing secretly in North Korea to gather information. They helped make raids to weaken the enemy supply chain. Working with the navy, they traveled as far as 40 miles (64 km) inland.

After the Korean War, ARC marines began training in more **unconventional** warfare tactics. They learned to parachute from airplanes and land safely in order to gather intelligence. At this point, ARC was made up of the Parachute Recon Platoon, the Pathfinder Platoon, and the Amphibious Recon Platoon. In 1957, these units joined to form the 1st Force Recon Company.

OORAH!

THE MARINE CRY "OORAH!" ORIGINATED WITH ARC MARINES DURING THE KOREAN WAR. AMPHIBIOUS MEMBERS TRAVELED ABOARD NAVY SUBMARINES. WHEN DIVING, THESE SUBS USED A SIGNAL THAT SOUNDED LIKE AN OLD-FASHIONED CAR HORN: AH-ROO-GAH! OVER THE YEARS THIS CHANGED TO OORAH!

AMERICAN SOLDIERS PARACHUTE INTO NORTH KOREA DURING THE KOREAN WAR.

VIETNAM AND BEYOND

In June 1958, Force Recon was expanded to include the 2nd Force Recon Company. In 1965, when the United States entered the Vietnam War, 1st Force Recon was sent to Vietnam, and 2nd Force Recon trained new Recon marines to go fight in Southeast Asia.

The Vietnam War was different from any war Force Recon marines had fought in before. The enemy was hard to find. Though their first task was reconnaissance, they also lured the enemy out of hiding and fought more often than before. In 1965, the 3rd Force Recon Company was established to help with reconnaissance in Vietnam.

CONTINUING RECON

FORCE RECON PLAYED AN IMPORTANT ROLE IN U.S. OPERATIONS IN THE PERSIAN GULF DURING THE EARLY 1990S. THEY WERE AN IMPORTANT FORCE IN THE WAR ON TERRORISM IN THE EARLY 2000S. THEY HAVE ALSO HELPED BATTLE SOMALI PIRATES IN THE GULF OF ADEN.

ON SEPTEMBER 8, 2010, SOMALI PIRATES CAPTURED THE GERMAN CARGO SHIP *MAGELLAN STAR*. ON SEPTEMBER 9, FORCE RECON MARINES JOINED OTHER SPECIAL FORCES TO HELP RESCUE THE 11 CREWMEMBERS AND CAPTURE THE 9 PIRATES. NO ONE WAS HURT.

RECON TOOLS

When Force Recon first formed during World War II, they had very basic tools to work with. Recon marines sometimes carried only knives to defend themselves. While an updated version of that KA-BAR knife is still used by marines, today's Force Recon units have the latest technology to aid them in their missions.

Because their job requires them to act like spies, Force Recon marines need high-tech surveillance gear. They use global positioning systems (GPS), parachutes, and scuba gear. The AN/PSQ-20 Enhanced Night Vision Goggle (ENVG) helps them see in the dark and can be held in the hand or mounted onto a helmet.

MODERN TECHNOLOGY HELPS FORCE RECON MARINES TO CARRY OUT DANGEROUS MISSIONS SUCCESSFULLY.

MODERN FORCE RECON

Ever since the terrorist attacks on the United States that took place on September 11, 2001, the U.S. military's main focus has been fighting terrorism. Marine Force Recon units are an important part of U.S. efforts to find and break up terrorist groups in Iraq, Afghanistan, and other countries.

Force Recon marines still spend time gathering information to locate enemy hiding places. However, their methods are changing. For Force Recon marines serving in the Middle East, this means being out in the open and showing people they're there to help. They spend more time listening to local citizens, rather than observing in secret.

THE NAME "BLACK DIAMOND" COMES FROM THE DIAMOND-SHAPED GOGGLE MOUNTS ON THEIR HELMETS.

DON'T MESS WITH THE BLACK DIAMONDS!

IN 2010, THE 1ST RECON BATTALION SPENT SEVEN MONTHS FIGHTING IN AFGHANISTAN WITHOUT LOSING ONE MAN TO ENEMY FIRE. THEY WORKED TO FIND TERRORISTS AND KEEP AFGHAN CITIZENS SAFE. THESE BRAVE SOLDIERS WERE KNOWN BY THEIR ENEMIES AS THE "BLACK DIAMONDS" BECAUSE OF A SHAPE ON THEIR HELMETS. FORCE RECON COMMANDERS HEARD TALIBAN SOLDIERS SPEAKING OVER THE RADIO: "DON'T MESS WITH THE BLACK DIAMONDS."

Supporting other military units has always been the main focus for Force Recon marines. In 2006, some Force Recon marines became part of the U.S. Marine Corps Special Operations Command (MARSOC), a new grouping within the U.S. Special Operations Command. The reorganization put Force Recon in the company of other specially trained fighting units from all branches of the armed forces.

Force Recon marines continue their covert reconnaissance missions all over the world. The information they gather is valuable. Today, reaching out to local people has become as important to Force Recon as getting intelligence. Force Recon continues to protect people all over the world.

A MARINE IN AFGHANISTAN RECEIVES ORDERS FOR HIS TEAM IN THS PHOTOGRAPH FROM 2011.

GLOSSARY

amphibious: Having the ability to operate both on land and in water.

battalion: A large organized group of soldiers.

demolition: The act of working with explosives to destroy things.

elite: Superior in talent and ability.

hostile: Unfriendly or aggressive.

marksmanship: Being skilled in shooting a gun at a target.

motto: A short expression of a guiding rule of conduct.

navigation: The science of figuring out the position and course of a ship or aircraft.

recruit: A new member of a military force. Also, to invite people to join a military force.

reserve: A military force that is additional to the regular forces and that is available if it is needed.

seize: To take by force.

sniping: Shooting well from a hiding place.

suspension: The act of hanging from a rope.

unconventional: Not conventional, or being out of the ordinary.

FOR MORE INFORMATION

BOOKS

Slater, Lee. *Marine Force Recon*. Minneapolis, MN: ABDO, 2015.

Terrell, Brandon. *Marine Force Recon*. Mankato, MN: The Child's World, 2018.

WEBSITES

Marines

www.marines.com

Learn more about what the marines and their Special Ops forces do.

National Museum of the Marine Corps

www.usmcmuseum.com

Visit this website to take a virtual tour of the National Museum of the Marine Corps.

Publisher's note to educators and parents: Our editors have carefully reviewed these websites to ensure that they are suitable for students. Many websites change frequently, however, and we cannot guarantee that a site's future contents will continue to meet our high standards of quality and educational value. Be advised that students should be closely supervised whenever they access the internet.

INDEX